TECHNICAL
REPORT

Monitoring Employment Conditions of Military Spouses

Nelson Lim, Daniela Golinelli

Prepared for the Office of the Secretary of Defense

Approved for public release; distribution unlimited

 NATIONAL DEFENSE RESEARCH INSTITUTE

The research described in this report was prepared for the Office of the Secretary of Defense (OSD). The research was conducted in the RAND National Defense Research Institute, a federally funded research and development center sponsored by the OSD, the Joint Staff, the Unified Combatant Commands, the Department of the Navy, the Marine Corps, the defense agencies, and the defense Intelligence Community under Contract DASW01-01-C-0004.

Library of Congress Cataloging-in-Publication Data is available for this publication.

ISBN 978-0-8330-3961-3

The RAND Corporation is a nonprofit research organization providing objective analysis and effective solutions that address the challenges facing the public and private sectors around the world. RAND's publications do not necessarily reflect the opinions of its research clients and sponsors.

RAND® is a registered trademark.

© Copyright 2006 RAND Corporation

All rights reserved. No part of this book may be reproduced in any form by any electronic or mechanical means (including photocopying, recording, or information storage and retrieval) without permission in writing from RAND.

Published 2006 by the RAND Corporation
1776 Main Street, P.O. Box 2138, Santa Monica, CA 90407-2138
1200 South Hayes Street, Arlington, VA 22202-5050
4570 Fifth Avenue, Suite 600, Pittsburgh, PA 15213
RAND URL: http://www.rand.org/
To order RAND documents or to obtain additional information, contact
Distribution Services: Telephone: (310) 451-7002;
Fax: (310) 451-6915; Email: order@rand.org

Preface

Decisions to enlist and especially to remain in the military are often not made alone. Most of the 1.4 million active-duty members are married, and they must consider the interests of their spouses when deciding whether or not to remain in the military. Spouses, in turn, will be influenced by their own civilian opportunities as circumscribed by the members' military life. Most spouses of active-duty personnel are active in the labor market (i.e., employed or seeking work). This status means civilian employment opportunities will affect how members view military life and how willing they are to continue committing their families to military life. The issues of military spouses in the labor market are therefore critical ones for retention of military members.

As part of the continuing research on the quality of life of military families, the RAND Corporation was asked to explore how to optimally use current data sources to monitor employment conditions of military spouses, and to determine if additional data sources are necessary. This document presents our research on developing employment statistics for military spouses.

This research was sponsored by the Military Families and Community Policy Office in the Office of the Secretary of Defense and conducted within the Forces and Resources Policy Center of the RAND National Defense Research Institute, a federally funded research and development center sponsored by the Office of the Secretary of Defense, the Joint Staff, the Unified Combatant Commands, the Department of the Navy, the Marine Corps, the defense agencies, and the defense Intelligence Community. Nelson Lim served as the principal investigator. Comments are welcome and may be sent to Nelson Lim at Nelson_Lim@rand.org.

For more information on RAND's Forces and Resources Policy Center, contact the Director, Jim Hosek. He can be reached by email at James_Hosek@rand.org; by phone at 310-393-0411, extension 7183; or by mail at the RAND Corporation, 1776 Main Street, Santa Monica, California 90407-2138. More information about RAND is available at www.rand.org.

Contents

Figures

Tables

Summary

The military community is largely one of families. Most active-duty personnel are married, and nearly half have children. Decisions to remain in the military are therefore influenced by how a military member views family life issues, including employment prospects for a member's spouse.

Most military spouses are active in the labor market. Military spouses, however, confront substantial obstacles to finding work and developing their own careers. Their unemployment rates are higher than those for civilian counterparts, and most military spouses perceive that being a military spouse adversely affects their work opportunities.

Given these issues, the Department of Defense (DoD) asked RAND researchers to help develop reliable employment statistics for military spouses. Developing those statistics would require determining (1) valid measures of labor market conditions for military spouses and (2) a sample of sufficient size to allow generalization to the population of military spouses. It is the purpose of this study to make those determinations.

Traditional measures established by the Bureau of Labor Statistics (BLS) divide the population into those who are in or out of the labor force. Those in the labor force are divided into full-time, part-time, and unemployed (but seeking work) workers; those outside the labor force are divided into persons voluntarily outside (e.g., retired) or involuntarily so (e.g., discouraged job-seekers). Additional BLS measures focus on varying levels of labor underutilization, ranging from the long-term unemployed to persons who are working part-time because full-time work is not available. All these measures have been tested over time through the Current Population Survey (CPS) and could be adapted to analysis of military spouses. In fact, the Status of Forces Survey (SOFS) of Active-Duty Members currently contains CPS questions to compute the BLS measures for military spouses.

The traditional measures, however, have at least two shortcomings for analysis of military spouses. First, they are based on activities in a given week. Because military families move often, a certain number of military spouses will be replying that they have no interest in the labor market, when in fact, they are simply too busy with moving and thus are between periods of job searching. Second, the traditional measures are designed to assess labor force conditions throughout the course of a business cycle. By contrast, DoD seeks measures that help describe economic hardships associated with the labor force status of military spouses.

Moreover, to craft policy decisions, DoD is interested in more than just whether military spouses are more or less likely to be employed or unemployed compared with their civilian counterparts. DoD needs to better understand why military spouses are less likely to be in the labor force and why they are more likely to be unemployed or underemployed. Employment and unemployment statistics alone do not help policy analysts to examine the underlying causes of relatively low employment or high unemployment. A richer characterization of labor market experiences is necessary to illuminate the possible causes that lead to different outcomes between military and civilian spouses.

Labor Utilization Framework (LUF) measures help better describe labor force status associated with economic hardship. Such measures classify workers by whether they are active or inactive, whether workers may be overqualified for the job they hold, whether they receive fewer hours than they desire, or whether they receive lower wages than are adequate. We recommend DoD develop and monitor both basic BLS and LUF indicators for military spouses.

Because such measures can be calculated through the use of the CPS, it is logical to consider whether the CPS can be used to devise measures for military spouses that are comparable to those for other civilian populations. This requires considering (1) whether the CPS includes a sufficient number of military families in its sample design, and (2) if not, whether it can still be adapted to be an efficient and cost-effective means of collecting labor force data on military spouses.

Unfortunately, the CPS is not likely to be adequate for analysis of military spouses. There are typically fewer than 500 military spouses in the CPS, a number that suffices only for measuring changes in military spouse employment or unemployment rates over the course of time that are of substantial magnitude (5 percentage points or more). Similarly, such a sample is adequate only for detecting employment or unemployment rate differences of similar magnitude between military and civilian spouses. Furthermore, the CPS, designed to assess the civilian population, systematically excludes areas such as military bases where many military families live. Redesigning the CPS to include more military families would likely carry prohibitively high costs.

Rather than collecting information on military spouses through other surveys, DoD can collect such information directly, and more efficiently, on its own. DoD could, for example, add questions to the Status of Forces Survey of Active-Duty Members or increase the frequency of the Surveys of Active-Duty Personnel and Spouses. Alternatively, given that these surveys are already quite lengthy, it could launch a new annual survey of military spouses. Such a survey could, for low cost, draw a representative sample of military spouses by using DoD's administrative databases. A sample of 5,000 to 10,000 military spouses would allow analysts to detect all but the smallest changes in labor force status for this population, though larger samples may be needed to analyze some specific subpopulations.

Acknowledgments

We are grateful to Aggie Buyers and Jane Burke of our sponsoring office for their support and assistance provided during this research. This research benefited from the assistance and intellectual contributions of many RAND colleagues, including Meg Harrell, Greg Ridgeway, Michelle Cho, Clifford Grammich, James Chiesa, Catherine Chao, Lizeth Bejarano, David Loughran, and Claude Messan Setodji.

Abbreviations

BLS	Bureau of Labor Statistics
CPS	Current Population Survey
DoD	Department of Defense
LUF	Labor Utilization Framework
NDRI	RAND National Defense Research Institute
PSU	primary sampling unit
SOFS	Status of Forces Survey
USU	ultimate sampling unit

CHAPTER ONE

The Need for Data on Military Spouses

The military community is largely one of families. Most active-duty personnel are married, and nearly half have children (Office of the Deputy Under Secretary of Defense for Military Community and Family Policy, 2004). Decisions to remain in the military are therefore influenced by how a military member views family life issues, including employment prospects for the member's spouse. Such issues become particularly important after initial years of service, when both the likelihood of the member starting a family and the investment that the military has made in the member are higher.

Most military spouses are active in the labor market, with 69 percent either employed or actively seeking work (Office of the Deputy Under Secretary of Defense for Military Community and Family Policy, 2004). Spouses' employment is an essential source of income for most military families; 68 percent of spouses work to help pay basic family expenses (Office of the Deputy Under Secretary of Defense for Military Community and Family Policy, 2004).

Military spouses, however, confront substantial obstacles to finding work and developing their own careers. Military spouses are less likely to be in the labor force. If they are in the labor force, they are more likely to be unemployed (or actively seeking work) and less likely to be employed than their civilian counterparts. Among the employed spouses, military spouses earn less than civilian spouses (Harrell et al., 2004). Most military spouses perceive that being a military spouse negatively affects their work opportunities (Harrell et al., 2004).

There are several explanations for the labor market disparities between military and civilian spouses. For example, the frequent (and often long-distance) moves associated with military life can be a main source of disruption to military spouses' employment. Alternatively, military spouses simply may have lower "taste" for work (Hosek et al., 2002). The "frequent moves" explanation entails very different policy prescriptions from the "low taste for work" explanation.

Given the importance of labor market conditions for military spouses, their members, and their families, as well as the challenges military spouses confront in the labor market, the Department of Defense (DoD) asked RAND researchers to help develop reliable employment statistics for military spouses on a continuing basis. Such statistics would enable DoD to monitor labor market conditions of military spouses over time, compare their conditions with those of civilian spouses, and develop effective policy interventions to alleviate difficulties faced by military spouses.

Because locating all persons in a population and recording their workforce data would be an enormous undertaking, labor market analysts rely on surveys of a sample of the population to generate reliable statistics. Reliable employment statistics for military spouses require

1. valid measures that appropriately capture labor market conditions of military spouses, and
2. a probability sample with sufficient size to allow generalization to the population of military spouses and to detect changes in their labor market conditions.

To monitor civilian labor force statistics, including the unemployment rate and those on other labor market issues, the Census Bureau and the Bureau of Labor Statistics (BLS) use the Current Population Survey (CPS). The CPS uses a stratified sampling to obtain a representative sample of the civilian noninstitutional population within the United States (Bureau of Labor Statistics, 2002). The sample contains about 50,000 housing units. Within each sampled unit, one person, who is at least 16 years old and not a member of the Armed Forces, is randomly selected as the reference person. Each month, an interviewer contacts the reference person to collect basic demographic information about all persons residing at the sampled units and detailed labor force information for all persons age 15 or over. Hence, it is logical to consider using the CPS to monitor changes in labor market situations of military families, and specifically the labor force status of military spouses.

Does the CPS have valid measures of workforce status as well as a sufficient sample of military spouses for analysis of their labor market conditions? If not, what alternatives might DoD consider? We explore these questions in this study. In Chapter Two, we identify appropriate labor status measures for military spouses and discuss whether these might be calculated from the CPS. In Chapter Three, we calculate how many military families would need to be included in a representative sample for estimating these measures. In Chapter Four, we conclude by qualitatively assessing the costs and benefits of approaches to remedy the shortfalls we find in the CPS sample, including expanding the CPS or fielding a special DoD survey.

CHAPTER TWO

Measures of Employment Conditions

The BLS has published monthly statistics on U.S. civilian labor market conditions since the 1940s. Over the years, the BLS has refined its measures for accuracy and validity. These measures are therefore logical candidates for DoD to consider in monitoring the labor market situations of military spouses.

In this chapter, we review the BLS measures of labor market activities, evaluate whether these measures yield valid information, and suggest additional or alternative measures that DoD may use to accurately capture the employment status of military spouses. In doing so, we will answer the following two questions:

1. Do existing BLS measures adequately capture the labor market experiences of military spouses?
2. If not, can more useful measures be constructed from currently fielded CPS questions that could then be incorporated into other DoD surveys?

BLS Basic Measures of Employment Conditions

The BLS measures on labor force status are based on CPS questions regarding work activity in the previous week (Appendix A). The BLS basic measures of employment conditions categorize the U.S. civilian population based on their labor market activities (Figure 2.1).

At the highest level of aggregation, the BLS divides the U.S. civilian population into two groups: individuals who are *in the labor force* and those *not in the labor force*. Being in the labor force is not the same as having a job. A person can be classified as a labor force participant as long as he or she is employed or actively looking for work (Bureau of Labor Statistics, 2003). Hence, the civilian labor force consists of employed and unemployed persons. The *employed* workers are further divided into *full-time* (worked 35 hours or more a week) and *part-time* (worked less than 35 hours a week) workers (Bureau of Labor Statistics, 2003).

An *unemployed* person is not the same as a jobless person. BLS defines unemployed persons as "all persons who were not employed during the [CPS] reference week, but were available for work (excluding temporary illness), and had made specific efforts to find employment some time during the 4-week period ending with the reference week" (Bureau of Labor Statistics, 2003, p. 2).

Figure 2.1
BLS Categories of Labor Market Conditions

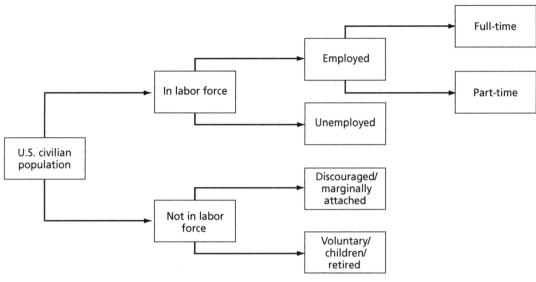

RAND *TR324-2.1*

Civilians who are not in the labor force can be divided into two groups: (1) *discouraged* and *marginally attached workers* and (2) other individuals who may be considered to be outside the labor force voluntarily or for other reasons, e.g., disabilities. Among persons voluntarily not in the labor force are children, retired persons, and persons pursuing other activities (e.g., education, child-rearing).

Discouraged workers are those who want work, are available for a job, and have looked for work sometime in the past 12 months, but are not currently looking because they believe there are no jobs available or there are none for which they would qualify (Bureau of Labor Statistics, 2003). *Marginally attached workers* are those who sought employment in the past 12 months, are available for and want work, but are not currently seeking employment because of child care or other family, school, and transportation hurdles.

The BLS uses the CPS to estimate the number of Americans in each group. For example, in March 2004,[1] the total number of persons at least 16 years of age in the civilian non-institutional population was 222.6 million (Bureau of Labor Statistics, 2004, Table A-1). Of these, 146.5 million, or 66 percent, were in the labor force; 76.0 million, or 34 percent, were not (Figure 2.2). Among the 146.5 million labor force participants, 137.7 million (62 percent) were employed—112.8 million (51 percent of total population) full-time and 24.9 million

---

[1]   Throughout this report, we describe various measures of labor market conditions using a representative sample of the civilian population instead of using a subsample of married women. This reflects the primary purpose of this section of the report, which is to describe concepts and reasons behind the measures, not to provide substantive analysis of differences in labor market conditions of civilian and military spouses. In Chapter Three, we discuss the difficulties associated with using the CPS sample for that type of analysis. Hence, the choice of the survey year, 2004, is arbitrary for the discussion. And it is important to note that the labor statistics reported in this section will change, if we use a different survey year or include only married women in the computation.

Figure 2.2
BLS Estimates of Labor Market Categories Among U.S. Civilian Adults, March 2004

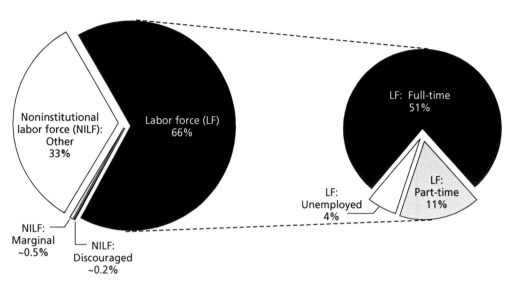

Total U.S. civilian population = 222.6 million (March 2004)

RAND *TR324-2.2*

part-time (11 percent)—and 8.8 million (4 percent of total population) were unemployed. Among those not in the labor force, about 500,000 (0.2 percent of the total) were discouraged workers and 1.1 million (0.5 percent) were marginally attached workers (Bureau of Labor Statistics, 2004, Table A-13).

Based on these population estimates, the BLS computes labor market statistics such as the civilian labor force participation rate, the employment-population ratio, and the unemployment rate. Mathematically, these statistics can be defined as follows:

$$\text{Labor Force Participation Rate} = \frac{\text{Labor Force}}{\text{Adult Civilian Population}}$$

$$\text{Employment–Population Ratio} = \frac{\text{Employed}}{\text{Adult Civilian Population}}$$

$$\text{Unemployment Rate} = \frac{\text{Unemployed}}{\text{Labor Force}}$$

Hence, in March 2004, the U.S. civilian labor force participation rate was 65.8 percent (146.5 million divided by 222.6 million), the employment-population ratio was 61.9 percent (137.7 million divided by 222.6 million), and the unemployment rate was 6.0 percent (8.8 million divided by 146.5 million).

Using BLS Basic Measures for Military Spouses

The BLS basic measures have been used for several decades and are well-established benchmarks for analyzing labor market conditions. Using them to assess labor market conditions of military spouses will yield measures that are comparable to those for their civilian counterparts. Fortunately, the Status of Forces Survey (SOFS) of Active-Duty Members in recent years has included a number of questions on spouses' labor market activities. These questions are identical to the CPS questions used by the BLS. As a result, DoD can use these questions to compute the labor force participation rate, employment-population ratio, and unemployment rate for military spouses, comparable to the rates published by the BLS for the adult civilian population.

Unfortunately, the labor market statistics that can be calculated from the CPS and the Status of Forces Survey of Active-Duty Members are based on questions regarding work-related and job search activities during a specific reference week. This limitation can distort labor market statistics for military spouses. More importantly, the labor market statistics based on the reference week would reveal little information about the underlying reasons military spouses are worse off than civilian spouses. For example, one of the main features of military life is frequent relocation. The moves, especially Do-It-Yourself Moves, rely heavily on the spouses' time to accomplish the move and thereby prevent spouses from looking for work just before and after the move (Hosek et al., 2002). If the reference week happens to fall within the time of moving activities, military spouses will naturally be classified as not engaged in labor market activities, as they may not be actively seeking work. While it is true that, on average, a certain percentage of military spouses are engaged in relocation and thus out of the labor force, that number overstates the percentage that are out of the labor force on any sustained basis.

The Status of Forces Survey of Active-Duty Members also does not contain CPS questions permitting analysis of discouraged and marginally attached workers among military spouses. These categories may account for higher percentages of military spouses than of civilians. Because their husbands (or wives) may be constrained by the call of duty from sharing child care responsibilities or providing emotional support and companionship, military spouses may not be as free to participate in the labor market as civilians. Military spouses also must participate in local labor markets that may not reward skills they have developed elsewhere or with which they may be unfamiliar or have few personal connections (Harrell et al., 2004).

BLS Alternative Measures of Labor Underutilization

The limitations of BLS basic measures, especially the official unemployment rate, have led the BLS to devise additional measures beyond those most frequently used and reported. The limitations of the official unemployment rate for assessing economic hardship, for example, led the BLS in 1976 to introduce a range of measures of labor underutilization, in addition to the official unemployment rate (Bregger and Haugen, 1995). These alternative measures of labor underutilization were modified in 1994 as a part of the redesign of the CPS. These BLS alternative measures of labor force underutilization contain six measures, U-1 to U-6 (Table 2.1). Each month, the BLS publishes the estimates of these alternative labor underutilization statistics along with its official labor market statistics. Such data help present a more complete picture of conditions for the unemployed or underused workers. Figure 2.3 represents BLS estimates of these measures.

Table 2.1
Range of Unemployment Measures (U-1 to U-6)

Category/Name	Description
U-1	Persons unemployed 15 weeks or longer, as a percentage of the civilian labor force
U-2	Job losers and persons who completed temporary jobs, as a percentage of the civilian labor force
U-3	Total unemployed, as a percentage of the civilian labor force (official unemployment rate)
U-4	Total unemployed plus discouraged workers, as a percentage of the civilian labor force plus discouraged workers
U-5	Total unemployed, plus discouraged workers, plus all other marginally attached workers, as a percentage of the civilian labor force plus all marginally attached workers
U-6	Total unemployed, plus discouraged workers, plus all other marginally attached workers, plus all persons employed part-time for economic reasons, as a percentage of the civilian labor force plus all marginally attached workers

U-1 reports the percentage of the civilian labor force that has been unemployed for more than 15 weeks. It aims to measure the level of long-term unemployment in the target population. This measure is the most restrictive measure of labor force underutilization and excludes those who may be considered unemployed in U-3, the official measure of unemployment. In March 2004, the BLS reported that 2.6 percent of the civilian labor force were unemployed persons who had been out of work at least 15 weeks.

Figure 2.3
BLS Estimates of Alternative Measures of Labor Underutilization, March 2004

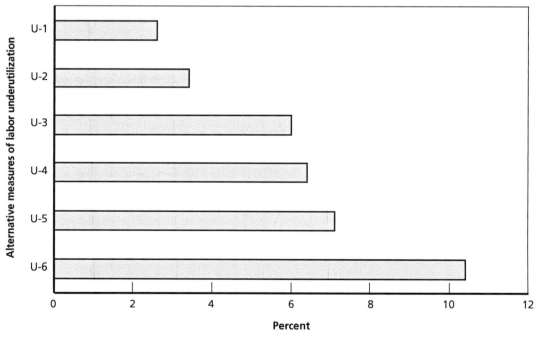

U-2 reflects the fraction of the civilian labor force that has recently lost permanent jobs or completed temporary jobs or that is on temporary layoff. In March 2004, the BLS found that 3.4 percent of the civilian labor force fell into this category.

U-3 is the BLS official measure of unemployment, which, as defined in the previous section, was 6.0 percent in March 2004.

U-4 and U-5 account for the prevalence of discouraged and marginally attached workers in the target population. In March 2004, there were 8.834 million unemployed persons and 514,000 additional persons not looking for work because of their discouragement over their job prospects; dividing the 9.348 million persons in these two categories by the sum of the number of persons in the civilian labor force and the number of discouraged workers yields a U-4 value of 6.4 percent. In addition to the persons who were unemployed or too discouraged to seek work, there were 1.130 million persons who had searched for work in the previous 12 months but were not searching for work now; dividing the sum of the unemployed, discouraged workers, and persons otherwise marginally attached to the labor force by the sum of the civilian labor force and the discouraged and marginally attached workers yields a U-5 value of 7.1 percent.

U-6 measures the level of labor underutilization by including involuntary part-time workers (i.e., workers who find only part-time jobs for economic reasons such as slack/bad conditions in business). It is the most inclusive measure of underutilization in the series, as it "effectively treat[s] workers who are visibly underemployed and all persons who are 'marginally attached' to the labor force equally with the unemployed" (Bregger and Haugen, 1995, p. 24). In March 2004, the CPS found 15.35 million persons were unemployed, working part-time for economic reasons, too discouraged to look for work, or marginally attached to the labor force; dividing this number by the sum of persons in the civilian labor force and the number of persons marginally attached to the civilian labor force yields a U-6 value of 10.4 percent.

Using BLS Alternative Measures of Labor Underutilization for Military Spouses

Alternative measures can help document labor underutilization among military spouses. In particular, U-4, U-5, and U-6, with their focus on persons in and outside the labor force who would like to find more work, can help document some of the unique circumstances military spouses confront in seeking work. Yet a principal aim of DoD in developing measures of employment conditions for military spouses is to measure the economic hardship experienced by military spouses and their families. Measures of economic hardship require information on income and the type of work done earning that income. We turn to additional measures DoD may use to assess employment conditions of military spouses.

Measures of the Labor Utilization Framework (LUF)

In addition to issues of little or no employment prospects, military spouses must typically also negotiate labor markets in areas that may have fewer or less suitable employment opportunities. For example, many military installations are in rural areas; assuming that jobs in the rural areas do not require a high level of education, military spouses, who have relatively high education levels, may only find jobs for which they are overqualified. Given local labor market conditions and circumstances of military life, they may be able to negotiate only part-time positions while desiring full-time work. Moreover, basing labor force measures on a single "reference week" may also lead to a distorted portrayal of the work experience of military spouses whose work experiences, because military families move often, are more likely to change week to week than those of comparable civilian spouses.

A labor force classification for military spouses should seek to document potential *underemployment* conditions such as "involuntary part-time work, *'working poverty,'* the gross misfit between educational credentials and employment opportunities, and discouragement" (Clogg et al., 1986, p. 118). Measures of the LUF, developed by researchers of social stratification, are designed in response to criticisms of traditional unemployment measures. These measures can portray such conditions better than the traditional BLS measures. The CPS in recent years has included all the necessary questions to develop LUF measures. In other words, field-tested survey questions are readily available for DoD to use to gather necessary information from military spouses.

The LUF measures aim to capture three dimensions of underemployment: "work time lost, income deficiency, and the mismatch of workers' skill attainment with required job skills" (Clogg et al., 2001, p. 20). The LUF uses two dimensions to represent a person's labor market situation: labor force behavior and labor force position.

LUF Measures of Labor Force Behavior

Labor force behavior is measured through all labor force activities of the past year, not a single week, to classify persons into 10 mutually exclusive categories[2] (Table 2.2). Labor force behavior measures account for both part- and full-time employment, whether or not a person is seeking employment, and the amount of time spent seeking employment. Combining these three elements provides a rich portrayal of a person's labor market situation during the previous year. The ten categories can be aggregated into three labor force statuses: (1) stable-inactive, i.e., nonworkers not looking for work; (2) stable-active, i.e., full-time, full-year workers; and (3) unstable-active, including all other categories of workers, e.g., persons at work who also sought work in the past year.

---

[2] These categories can be expanded to 16 categories, if including the number of stretches in which a person looked for work (Clogg et al., 2001, p. 121). Such detail is not needed for measuring labor market employment for military spouses. Attaining such detail would require collecting additional information because stretches of job searching may be confounded with the periods of relocation.

Table 2.2
Categories of Labor Force Behavior (Past Year)

Category/Name		Description
Stable-inactive		
1	NW-NL	Nonworker
Unstable-active		
2	PTPY-NL	Part-time, part-year worker, not looking
3	PTPY[15+]	Part-time, part-year worker, looked 15+ weeks
4	PTPY[14]	Part-time, part-year worker, looked 1–14 weeks
5	FTPY-NL	Full-time, part-year worker, not looking
6	FTPY[15+]	Full-time, part-year worker, looked 15+ weeks
7	FTPY[14]	Full-time, part-year worker, looked 1–14 weeks
8	PTFY-OTHER	Part-time, full-year worker, voluntary
9	PTFY-INVOL	Part-time, full-year worker, involuntary
Stable full-time active		
10	FTFY	Full-time, full-year worker

LUF Measures of Labor Force Position

Labor force position is measured through eight categories showing employment status in the reference week, including hours worked, occupation, and earnings (Table 2.3). Some of the categories are similar to the BLS measures, but there are important differences, particularly among employed persons. The LUF groups employed persons into five different categories. Part-time workers are divided into two groups based on whether they voluntarily choose to work part-time or do so because full-time work is unavailable. Full-time workers are grouped into three groups: working poor (i.e., those who do not earn an annual wage of at least 125 percent of the poverty threshold[3]), overqualified individuals who work in positions requiring relatively low skills, and fully employed workers with adequate income.

LUF Statistics for March 2004

Unlike the BLS measures of labor market conditions we reviewed above, LUF measures are not readily available, but the CPS contains all information necessary to calculate them. For example, between March 2003 and March 2004, 56 percent of civilian adults, 16 years or older, were full-time, full-year workers; 18 percent were nonworkers; and 27 percent were active in the labor market but had unstable work (Figure 2.4).

Nearly half, 44 percent, of civilian adults were employed with adequate income in March 2004; i.e., they had full-time jobs with incomes above 125 percent of the poverty threshold (Figure 2.5) and were not educationally mismatched. As already mentioned, about one in five, 18 percent, was not in the labor force. Among the remaining labor market position categories,

---

[3]   For details on how these thresholds are calculated, see Clogg et al., 2001, especially pp. 37–39.

Table 2.3
Categories of Labor Force Position (Reference Week)

Category/Name		Description
1	NW-NL	Not in the labor force—economically inactive and not seeking employment in the labor market
2	S-U	Sub-unemployed—discouraged and conditionally interested workers in the CPS definition
3	U	Unemployed—the same as the CPS definition
4	H-I	Part-time employed (low hours)—involuntary
5	H-V	Part-time employed (low hours)—voluntary
6	I	Underemployed by low income (earnings)—employment that does not provide annual income greater than or equal to 125 percent of the poverty threshold
7	M	Educational mismatch—possession of educational qualifications (years of education) that are more than one standard deviation above the occupational average
8	Adequate Full-Time	Full-time workers with adequate income—residual category

Figure 2.4
LUF Labor Market Behavior Categories Among Age 16 or Older U.S. Civilians, March 2004

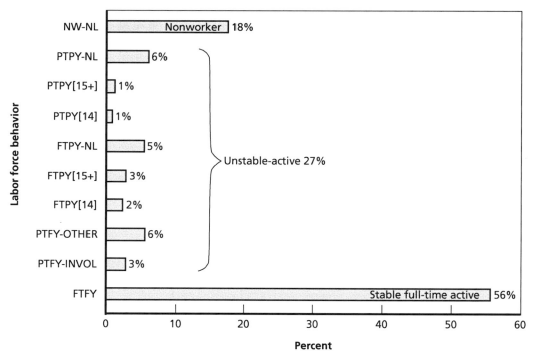

Figure 2.5
LUF Labor Market Position Categories Among U.S. Civilian Adults, March 2004

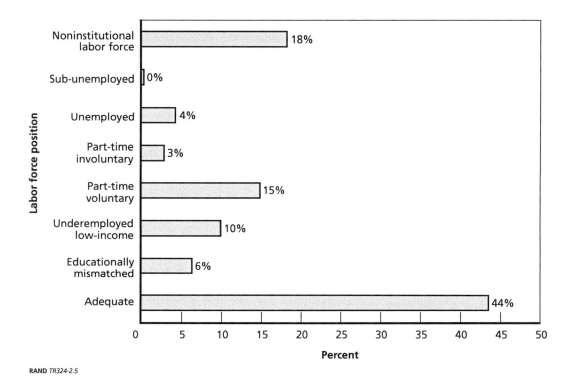

RAND *TR324-2.5*

we note in particular 6 percent who had an educational mismatch, i.e., educational qualifications exceeding those required for their job; 10 percent who were working poor, i.e., with employment that did not provide a wage exceeding 125 percent of the poverty threshold; and 3 percent who were working part-time involuntarily.

Using LUF for Military Spouses

The main advantage of the LUF measures in studying the labor market situation of military spouses is their portrayal of more-varied work experiences as well as of activities over the previous year rather than a single week. Such measures would help portray more detail on military spouses not at work, especially those who may be too discouraged to seek work or who may be marginally interested in employment, as well as those in the midst of relocation. The LUF measures also help gauge underemployment among military spouses, whether related to working part-time involuntarily, earning wages below the poverty threshold, or working in jobs for which one is overqualified, conditions which may affect military spouses more than other civilians. DoD should develop and monitor LUF indicators for military spouses. The current CPS instrument contains all the necessary survey questions to do so. In the next chapter we discuss the most efficient means DoD may use to collect the requisite data.

Summary

At the beginning of this chapter, we posed two questions on CPS measures of employment. Here are the answers we found.

Do traditional BLS measures adequately capture the labor market experiences of military spouses? No. Military spouses face unique circumstances such as frequent and long-distance moves that may result in labor market behavior and positions that traditional measures of civilian labor force activity do not adequately document.

If not, can more useful measures be constructed from currently fielded CPS questions that could then be incorporated into other DoD surveys? Yes. The BLS supplementary measures and the LUF measures derived from CPS data could help document military spouses who leave the labor market because of discouragement toward their job prospects as well as those who are underemployed. Though the BLS reports only few measures based on the CPS, the CPS can in fact be used to construct many more measures, including those for the LUF. In addition, CPS questions needed to collect necessary information can easily be incorporated into existing DoD surveys.

In sum, CPS questions suffice to document the labor force status of military spouses. We turn next to the question of the number of military families required to constitute a sufficient sample size that will allow generalization to the entire population of interest, and to detect small but significant changes in military spouses' labor market conditions.

CHAPTER THREE

Adequacy of the CPS Sample

To monitor civilian labor force statistics, including the unemployment rate and those on other labor market situations, the BLS uses the CPS. Could DoD use the CPS to devise measures of the employment conditions of military spouses? To do so, the CPS sample would have to include enough military families to allow inferences that are statistically valid. Thus, in this chapter, we answer two questions:

1. Does the CPS currently include a sufficient number of military families in its sample design?
2. If not, then is it the most efficient and cost-effective means of collecting labor force data on military spouses?

We begin with some background on sample adequacy from a statistical point of view. We then determine the number of military families in the sample that would be required to detect (a) policy-relevant changes in employment or unemployment rates from one year to the next and (b) policy-relevant differences in the employment or unemployment rates of military and civilian spouses. We compare these minimal sample sizes with the number of military families in the CPS and, finding the latter too small, consider whether the CPS might be modified to sufficiently increase the number of military families.

Adequacy of a Sample

The adequacy of a sample size depends on a variety of factors, including

- how small a change or difference one wishes to detect,
- how much uncertainty is associated with labor market statistics under examination, and
- how confident one wishes to be that the inference one draws is correct.

The first issue requires a decision regarding the importance of small changes in the employment of military spouses or small differences between their employment and that of civilians. Is a 1-percentage-point change or difference in employment rates enough to require a DoD intervention? Are 5 percentage points enough? Only DoD can make the decision as to when it should intervene. For the sake of determining adequacy of sample size, however, we must make some assumptions as to what would be a range of differences of interest to DoD. We thus assume here that DoD might be interested in changes or differences on the order of 1 to 5 percentage points.

The second issue is related to the uncertainty of estimates of labor market conditions of military spouses—such as the unemployment rate. The unemployment rate is estimated based on a representative sample of the population of military spouses. Since it is based on a sample, each estimate has a degree of uncertainty associated with it. This uncertainty is measured by the variance, which, in turn, affects the adequacy of a sample size. The greater the uncertainty is, the larger the variance is. The larger the variance is, the bigger the sample size needs to be to estimate the unemployment rate with desired precision.

Addressing the third issue requires some understanding of statistical hypothesis testing. The key decision in statistical testing is whether to accept or reject the *null hypothesis* that there is no change or difference. In the matter at hand, the null hypothesis is that there is no change in the employment rate from one year to the next or that the difference between the military and civilian spouses' employment rates is zero. The objective is to reject the null hypothesis when in fact it is wrong and to accept it when in fact it is right. There are two ways to make an erroneous decision: We can reject the null hypothesis when it is true (this is called a "type I error"). That is, we can infer there is a difference when none exists (called a "false positive" in some contexts). Or, we can accept the null hypothesis when, in reality, it is false (this is called a "type II error"). That is, we can miss a genuine difference by inferring there is none (a "false negative").

Given a certain sample size and other parameters, the probability of making a type I or type II error can be calculated. Or, more to the point, given an objective of holding type I and II errors in finding, say, a 2-percentage-point difference to some specified level, a sample size can be calculated. Typically, a test is constructed in such a way that, for a given value of the probability of making a "type I error" (designated α and usually set to 5 percent), it minimizes the probability of making a "type II error" (designated β).

If β represents the probability that, given a difference, we think there is none, then $(1 - \beta)$ represents the probability that, given a difference, we recognize that there is one. In the matter at hand, β represents the probability of erroneously concluding, for example, that the employment rate has not changed over time, when in reality it has, and $(1 - \beta)$ represents the probability of *correctly* concluding that, for example, the employment rate has changed over time. The latter probability, $1 - \beta$, is called the power of the test (Appendix B). We usually want a test that rejects the null hypothesis, when it is false (i.e., that recognizes true differences), with a high probability (i.e., a large power). A typical value for the power is 80 percent. In what follows, we will determine the sample size needed to detect changes in the employment rates that range from 1 to 5 percent, having set α equal to 5 percent and $(1 - \beta)$ equal to 80 percent. We will see that, under these settings, the smaller the change that we want to detect with a probability of 80 percent, the larger the sample size needs to be.

A Sample Size Adequate to Detect Changes from Year to Year

Here, we want to test the following null hypothesis:

The current military spouses' employment/unemployment rate is equal to the previous year's employment/unemployment rate.

Employment rates tend to be much larger than unemployment rates, and the sample size required depends not only on the factors already mentioned but also on the size of the numbers of interest. Larger sample sizes are required to detect differences in percentages close to 50, as employment rates can be, than to detect equal percentage-point differences in the smaller percentages characterizing unemployment rates.

Detecting Changes from a 13 Percent Baseline Unemployment Rate

Assuming that the unemployment rate in the previous year (or some other prior time) is 13 percent, we estimate the sample size necessary to detect changes of 5 percentage points or less. Figure 3.1 shows the relationship between the sample size, plotted along the y-axis, and the detectable changes from one year to the next, plotted along the x-axis. Since the new rate can be higher or lower than the historical rate (13 percent), there are two lines in the figure—one for positive differences and another for the negative ones.

The figure displays several features of the relationship. It shows that detecting smaller differences requires larger sample sizes. For example, to detect a 1-percentage-point change—meaning that the new unemployment rate is 12 percent or 14 percent—requires a sample size of about 10,000, but to be able to detect a 2-percentage-point change, a sample of about 2,000 suffices. In other words, with a sample of 2,000 Army wives, we can detect changes in the unemployment rate of at least 2 percentage points with a statistical power of 80 percent and with an α of 5 percent.

The benefit of including more spouses in the sample declines quickly. The first 2,000 spouses enable us to detect changes as small as 2 percentage points, while an additional 8,000 spouses are needed to be able to detect changes close to 1 percentage point.

Detecting Changes from a 50 Percent Baseline Employment Rate

Figure 3.2 displays the sample size required to detect changes when the prior rate is close to 50 percent, the employment rate for Army wives in 1989 (Harrell et al., 2004). Note that, compared with the previous example, a larger number of military spouses is needed to detect changes of any given size. In this example, about 5,000 spouses are needed to detect a 2-percentage-point change in the rate; in the previous example, 2,000 spouses were sufficient. As mentioned above, the larger the variance is, the bigger the sample size needs to be to estimate the unemployment rate with desired precision. The closer the baseline rate is to 50 percent, the

Figure 3.1
Sample Sizes Required to Detect Changes of 5 Percentage Points or Less from a Baseline Unemployment Rate of 13 Percent

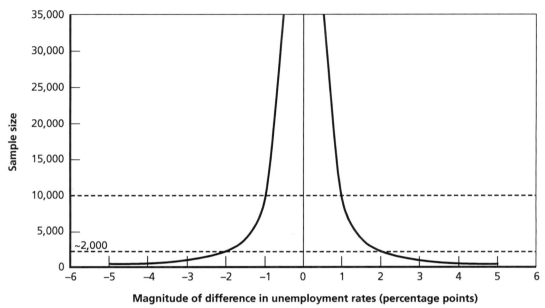

RAND *TR324-3.1*

Figure 3.2
Sample Sizes Required to Detect Changes of 5 Percentage Points or Less from the Baseline Employment Rate of 50 Percent

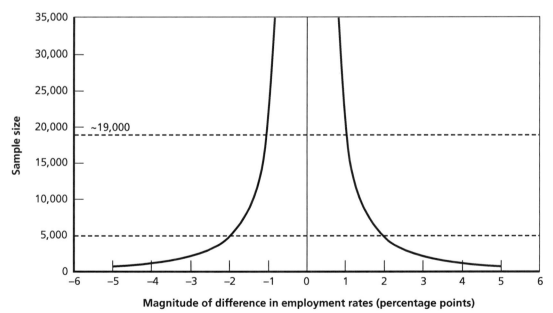

RAND *TR324-3.2*

larger the required sample size has to be. The reason is that the variance is largest in this setting. [The variance for a sample proportion (p) is

$$\left(\frac{p(1-p)}{n} \right).$$

For a given sample size, n, as p increases up to 50 percent, the variance increases; once p is greater than 50 percent, the variance declines. The shape of the relationship between the variance and the sample proportion is an upside-down U-shape with the peak at $p = 50$ percent.] Hence, the estimation of the sample size using 50 percent as the baseline rate provides the most conservative (largest) estimate of the sample size needed.

A Sample Size Adequate to Detect Military-Civilian Differences

Here, we want to test the following null hypothesis:

The military spouses' employment/unemployment rate is equal to the civilian spouses' rate.

We will assume that the sample size for the civilian spouses is larger than that for the military spouses. More specifically, we can assume that DoD will use the CPS sample of the civilian spouses. On average CPS collects data on 27,700 civilian families per year (Hosek et al., 2002). As CPS uses a complex sampling design, the *effective sample size* (i.e., the size of a simple random sample providing the same level of accuracy) for the civilian spouses sample is likely to be smaller. If we assume that because of the complex sampling design, there is a loss of 9 percent accuracy, the effective sample size will be approximately 25,000. If the loss in accuracy is 50 percent, which is not unreasonable for a design as complex as that used by the CPS, the effective sample size will be between 13,000 and 14,000 civilian spouses. To be more conservative, we conduct these two power analyses assuming that the civilian sample size is 10,000. As in the one-sample case, we specify the differences we wish to detect and determine the sample size of military spouses needed to detect such differences with a statistical power $(1 - \beta)$ of 80 percent and an α of 5 percent.

Detecting Differences from a 50 Percent Civilian Spouse Employment Rate

To be conservative, we consider the case in which the employment rate for the civilian spouses is 50 percent. Under this assumption, we determine the military spouse sample size for detecting differences between the employment rates of civilian and military spouses ranging from −5 to +5 percentage points with a statistical power of 80 percent and with an α of 5 percent. Results are reported in Figure 3.3.

The basic structure of Figure 3.3 is similar to the previous figures. However, there are a few notable differences. First, as the "baseline" in this case is an employment rate estimated

Figure 3.3
Sample Sizes Required to Detect Differences of ±5 Percentage Points or Less Between the Military and Civilian Spouses' Employment Rate, When the Civilian Employment Rate (p$_c$) Is 50 Percent and the Sample Size for the Civilian Spouses Is 10,000

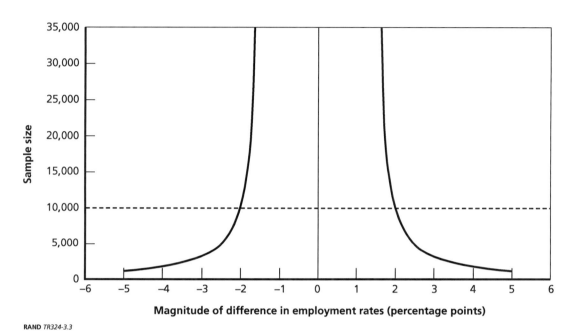

RAND *TR324-3.3*

from a sample with some uncertainty, larger numbers of military wives are needed to detect differences with the same confidence. It is not possible to detect very small differences, such as 1 percentage point, at all, unless the civilian wives sample size is increased. About 10,000 military spouses are needed to detect differences as low as 2 percentage points.

Detecting Differences from an 80 Percent Civilian Spouse Employment Rate

Though to be conservative, we have analyzed a 50 percent employment rate for civilian spouses, that rate is likely to exceed 50 percent. For example, Hosek et al. (2002) estimated that, between 1987 and 1999, the employment rate for civilian spouses with a high school education was around 80 percent. Figure 3.4 shows the sample size needed to detect a range of differences when the civilian spouse employment rate is 80 percent. The basic structure is similar to that shown in Figure 3.3, but smaller sample sizes are required to identify changes of equal magnitude. In this example, a sample of about 5,000 military spouses would be sufficient to detect changes as low as 2 percentage points. With a civilian sample of 10,000 respondents, however, it is again not possible to detect changes that are 1 percentage point or less with a statistical power of 80 percent and with an α of 5 percent. To do so would require increasing the civilian sample size and/or reducing the required level of power.

Figure 3.4
Sample Sizes Required to Detect Differences of ±5 Percentage Points or Less Between the Military and Civilian Spouses' Employment Rate, When the Civilian Employment Rate (p_c) Is 80 Percent and the Sample Size for the Civilian Spouses Is 10,000

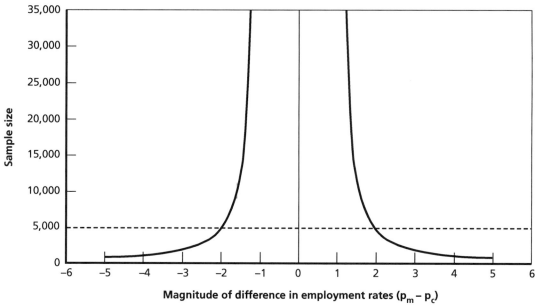

RAND *TR324-3.4*

Using CPS to Monitor Military Spouses' Employment Condition

We have shown that sample sizes of 2,000 military spouses or more would be required to detect small differences (2 percentage points) in employment or unemployment rates with good power (80 percent). In contrast, the number of military families in the CPS sample is quite small, about 448 per year on average (Hosek et al., 2002, p. 89). Given such a small number, only relatively large changes (5 percentage points or more) in labor market circumstances for military spouses are likely to be detected with such data.

To overcome the limitation of the small number of military families, one could pool multiple years of the CPS to generate a large sample of military spouses, as Hosek et al. (2002) had done. But this approach will not allow DoD to monitor annual changes of labor market conditions of military spouses.

Moreover, one of the reasons for the small number of military families in the CPS is that the current sampling strategy of the CPS effectively limits the inclusion of military families. The CPS seeks a representative sample of the civilian noninstitutional population within the United States. It uses a stratified sampling design[1] conducted in three stages:

---

[1] A stratified random sample is one "obtained by separating the population elements into nonoverlapping groups, called strata, and then selecting a simple random sample from each stratum" (Scheaffer et al., 1996, p. 125).

- First, the United States is divided into primary sampling units (PSUs), typically metropolitan areas, large counties, or groups of smaller counties. PSUs are grouped into strata using decennial census and other information, with each stratum having similar social and economic characteristics. One PSU is randomly selected per stratum.
- Second, a sample of housing units is drawn from the selected PSU. Because the CPS is designed to represent the civilian population, these "ultimate sampling units," or USUs, are never selected from military bases.
- Third, within households of each USU, one person, the reference person, selected from individuals who are at least 16 years of age and not in the Armed Forces, responds for all civilian members of the household.

The focus of the CPS on civilian households does not mean that military households are excluded. Military households are included in the CPS if they are off a military base. Hence, the military spouses that are included are no longer a *nationally* representative sample of the population of military spouses.

Could the CPS be modified to incorporate a greater number of military families? Including more military families would require adjustment of all stages of the sampling design to include military bases and other geographical areas in which military families reside. Substantial effort would be needed both to design a new sampling frame and to determine the number of military families that would be needed to produce stable estimates of labor market conditions of military spouses. The cost of making these changes would likely be high.[2]

Collecting data on military spouses, however, does not require a multistage design. DoD could easily select a simple random or stratified sample from the personnel databases. These databases provide readily available sampling frames that make sample selection pretty straightforward.

Summary

At the beginning of this chapter, we posed two questions on the sample design of the CPS and its possible uses for analyzing labor force circumstances among military spouses.

Does the CPS currently include a sufficient number of military families in its sample design? No. There are typically fewer than 500 military spouses in the CPS, a number that suffices only for detecting large differences.

If not, then is it the most efficient and cost-effective means of collecting labor force data on military spouses? No. The CPS instrument contains appropriate questions for tracking the experiences of military spouses in the labor market, and the CPS survey design could be adapted to changing labor market conditions. Nevertheless, the multistage stratified sampling design of the CPS is an inefficient way to survey the military population. Including more military families in the CPS would be quite expensive.

---

[2] Based on an informal interview with BLS personnel.

Rather than seeking to adapt the CPS sampling frame to its needs, DoD should consider using the CPS instrument (i.e., questionnaire) to gather data from DoD's own personnel databases. We discuss this in Chapter Four.

Conclusions and Recommendations

This research assessed various BLS measures used to monitor the changes in labor market conditions of military spouses compared with those of their civilian counterparts. We concluded that traditional and supplementary employment measures currently used by BLS may not be adequate to represent the impact of military life on military spouses' employment conditions. In order to fully capture the employment situations of military spouses, we recommend that DoD supplement BLS employment measures with measures derived from the LUF.

In addition, we conducted calculations to determine the sample size needed to detect with good power the temporal changes in the labor market conditions of military spouses, as well as differences in these conditions between military and civilian spouses. Based on this analysis, we concluded that the current number of military families in the CPS is not adequate for the stated goals. In addition, given the CPS's complex sampling design, we judged that increasing the number of military families in the CPS would be costly.

Rather than collecting information on military spouses through other surveys, DoD can collect such information directly, and more efficiently, on its own. It might, for example, add questions to existing DoD surveys (e.g., the Status of Forces Survey of Active-Duty Members) or increase the frequency of the Surveys of Active-Duty Personnel and Spouses. Both surveys have more than sufficient sample sizes. For instance, in 1999, the sample size for the Survey of Active-Duty Personnel was 66,040 and that for the Survey of Spouses of Active-Duty Personnel was 38,901. At the same time, however, the questionnaires for both these surveys are already quite lengthy and should not have still more questions added to them without deleting existing questions. The burden on service members to complete these surveys has serious consequences on the representativeness of the responses. In recent years, the response rates for these DoD surveys have been declining. For example, the response rate for the 2004 Status of Forces Survey (SOFS) of Active-Duty Members was 35.8 percent (Defense Manpower Data Center, 2004, p. 11), yet the response rate for the 1999 SOFS was 56.2 percent (Defense Manpower Data Center, 2000, p. 13).

A better solution might be for DoD to launch a new annual survey covering a substantially smaller number of military spouses. Such a survey could draw, for a relatively small cost, a representative sample of military spouses from the DoD's administrative databases, such as the Active-Duty Master File.

Gathering information directly from military spouses about their labor market activities will certainly improve accuracy of data. On the other hand, DoD should note that this strategy deviates from the CPS approach to collect similar information from civilian households. As we described above, the CPS gathers information about the adult members of a sampled household from a randomly selected member of the household. How these different approaches impact the comparability of the labor market statistics is not clear.

The new survey should contain demographic questions as well as 30 to 35 questions similar or identical to those in the CPS used to compute official BLS statistics. A sample of 2,000 to 3,000 military spouses would allow analysts to detect all the policy-relevant differences. It is important to note, however, that the magnitude of the policy-relevant differences cannot be determined based solely on technical statistical power calculations. As we stated at the beginning of Chapter Three, only DoD can determine the level of changes or differences in employment conditions that is enough to require a DoD intervention. For instance, if DoD decides that a change (or a difference) of 5 percentage points or more is needed to trigger a policy intervention, a sample of fewer than 1,000 military spouses will be sufficient. On the other hand, if DoD wants to study the employment conditions for particular subsets of the military spouses population, then a larger sample size would have to be considered.

Current Population Survey (CPS) Education and Employment Questions

Education

1. What is the highest level of school your spouse has completed or the highest degree he/she has received?

    a. Less than 1st grade
    b. 1st, 2nd, 3rd, or 4th grade
    c. 5th or 6th grade
    d. 7th or 8th grade
    e. 9th grade
    f. 10th grade
    g. 11th grade
    h. 12th grade (no diploma)
    i. High school graduate, high school diploma, or the equivalent (for example: GED)
    j. Some college but no degree
    k. Associate degree in college occupational/vocational program
    l. Associate degree in college academic program
    m. Bachelor's degree (for example: BA, AB, BS)
    n. Master's degree (for example: MA, MS, MEng, MEd, MSW, MBA)
    o. Professional school degree (for example: MD, DDS, DVM, LLB, JD)
    p. Doctorate degree (for example: PhD, EdD)

I am going to ask a few questions about work-related activities last week. By last week, I mean the week beginning on Sunday, [DATE], and ending on Saturday, [DATE].

Employment Activities

2. Last week, did your spouse do any work for pay?
    a. Yes
    b. No   (GO TO 5)
    c. Retired   (GO TO 20)
    d. Disabled   (GO TO 12)
    e. Unable to work   (GO TO 12)

3. How many hours per week does your spouse usually work at his/her job?
    – Number of hours _____

4. Last week, how many hours did your spouse actually work at his/her job?
    – Number of hours _____

    – (If less than 35, GO TO 9)

    – (If 35 or more, GO TO 20)

5. Last week, was your spouse on layoff from a job?
    a. Yes
    b. No (GO TO 12)

6. Has your spouse been given any indication that he/she will be recalled to work within the next 6 months?
    a. Yes
    b. No (GO TO 12)

7. Could he/she have returned to work last week if he/she had been recalled?
    a. Yes (GO TO 20)
    b. No (GO TO 8)

8. Why is that?
    a. Own temporary illness
    b. Going to school
    c. Other

GO TO 20

9. What is the main reason your spouse worked less than 35 hours last week?
    a.   Slack work/business conditions
    b.   Seasonal work
    c.   Job started or ended during week
    d.   Vacation/personal day
    e.   Own illness/injury/medical appointment
    f.   Holiday (legal or religious)
    g.   Child care problems
    h.   Other family/personal obligations
    i.   Labor dispute
    j.   Weather affected job
    k.   School/training
    l.   Civic duty
    m.   Other reason

10. Does your spouse want to work a full-time workweek of 35 hours or more per week?
    a.   Yes
    b.   No
    c.   Regular hours are full-time

11. What is your spouse's main reason for working part-time instead of full-time?
    a.   Slack work/business conditions
    b.   Could only find part-time work
    c.   Seasonal work
    d.   Child care problems
    e.   Other family/personal obligations
    f.   Health/medical limitations
    g.   School/training
    h.   Retired/Social Security limit on earnings
    i.   Full-time workweek is less than 35 hours
    j.   Other (specify)

GO TO 20

12. What was the main reason your spouse was absent from work last week?
    a.   Slack work/business conditions
    b.   Waiting for a new job to begin   (GO TO 20)
    c.   Vacation/personal days   (GO TO 20)
    d.   Own illness/injury/medical problems
    e.   Child care problems
    f.   Other family/personal obligation
    g.   Maternity/paternity leave   (GO TO 20)
    h.   Labor dispute
    i.   Weather affected job
    j.   School/training
    k.   Civic duty
    l.   Does not work in the business
    m.   Other (specify)

13. Has your spouse been doing anything to find work during the last four weeks?
    a.   Yes   (GO TO 15)
    b.   No

14. What is the main reason he/she was not looking for work during the last four weeks?
    a.   Believes no work available in line of work or area
    b.   Couldn't find any work
    c.   Lacks necessary schooling, training, skills, or experience
    d.   Employers think too young or too old
    e.   Other types of discrimination
    f.   Can't arrange child care
    g.   Family responsibilities
    h.   In school or other training
    i.   Ill health, physical disability
    j.   Transportation problems
    k.   Other (specify)
    l.   Completing move; settling in to new location
    m.   Anticipating move; did not want to commit to new job
    n.   Away while service member was deployed

15. As of the end of last week, how long had your spouse been looking for work?
    a.   Weeks
    b.   Months
    c.   Years

16. Exactly how many weeks has your spouse been looking for work?
    – Number of weeks _____

17. Has your spouse been looking for full-time work of 35 hours or more per week?
    a. Yes
    b. No
    c. Doesn't matter

18. Last week, could your spouse have started a job if one had been offered?
    a. Yes   (GO TO 20)
    b. No

19. Why is that?
    a. Waiting for new job to begin
    b. Own temporary illness
    c. Going to school
    d. Other (specify)

Characteristics of Work

Now I have a few questions about the job . . .
    a. at which your spouse worked last week
    b. from which your spouse was absent last week
    c. from which your spouse is on layoff
    d. at which your spouse last worked.

20. Was he/she employed by government, by a private company, a nonprofit organization, or was he/she self-employed or working in the family business?
    a. Government   (GO TO 21)
    b. Private for-profit company   (GO TO 22)
    c. Nonprofit organization including tax-exempt and charitable organizations (GO TO 22)
    d. Self-employed   (GO TO 22)
    e. Working in the family business   (GO TO 22)

21. Would that be the federal, state, or local government?
    a. Federal
    b. State
    c. Local (county, city, township)

GO TO 23

22. Is this business or organization mainly manufacturing, retail trade, wholesale trade, or something else?
    a. Manufacturing
    b. Retail trade
    c. Wholesale trade
    d. Something else

23. What kind of work does your spouse do; that is, what is his/her occupation? (For example: plumber, typist, farmer)

    — _____

24. What are his/her usual activities or duties at this job? (For example: typing, keeping account books, filing, selling cars, operating printing press, laying brick)

    — _____

25. How much did your spouse earn from all employers before taxes and other deductions during last year?

    — _____

If your spouse is self-employed:

26. How much did your spouse earn (net) from this business/farm after expenses during last year?

    — _____

Power Analysis

In this appendix, we describe technical details of how to determine an adequate sample size to detect specified differences in both the employment and unemployment rates with a given power under two scenarios. The results are discussed in the report.

As we described in the report, in the first scenario, we are interested in testing whether the employment and unemployment rates for the military spouses increase or decrease from one year to the next. In particular, we assume to know the previous year employment and unemployment rates and we determine how large the sample of military spouses has to be in order to detect prespecified changes in such rates with sufficient power. The first scenario is equivalent to conducting a one-sample test.

In the second scenario, we are interested in testing the difference in employment and unemployment rates between military and civilian spouses. Again, we want to determine the sample size necessary for the military spouses in order to be able to detect prespecified differences with sufficient power. The second scenario is equivalent to conducting a two-sample test.

In the computations described below, we assume that both the samples of military and civilian spouses are simple random samples. In other words, the determined sample sizes are effective sample sizes. If a complex design is adopted to obtain the samples, their nominal sample sizes will likely be larger than the sizes of the simple random samples in order to provide the same level of accuracy provided by the simple random samples.

Scenario 1: One-Sample Power Analysis

We want to test, using a two-sided test with an α-level of 5 percent:

$$H_0: p = p_0 \text{ versus } H_a: p \neq p_0,$$

where p_0 is the previous year employment/unemployment rate, which we assume known, and p is the current employment/unemployment rate.

We have a sample of n independent and identically distributed Bernoulli random variables with $P(X_i = 1) = p$. Given a sample of military spouses of size n, we can estimate the current employment (or unemployment) rate p using the sample proportion $\hat{p} = \dfrac{\sum_{i=1}^{n} x_i}{n}$. If the sample size n is large enough, then the sample proportion will be approximately distributed as normal with mean p and variance $p(1-p)/n$.

The α-level will determine how large such difference has to be in order to reject H_0, when H_0 is true:

$$\alpha = P\left(\left|\frac{\hat{p} - p_0}{\sqrt{\frac{p_0(1-p_0)}{n}}}\right| > z_{\alpha/2}\right) = P\left(\hat{p} > z_{\alpha/2}\sqrt{\frac{p_0(1-p_0)}{n}} + p_0\right) + P\left(\hat{p} < -z_{\alpha/2}\sqrt{\frac{p_0(1-p_0)}{n}} + p_0\right),$$

where z is the quantile of a standard normal.

The power $(1 - \beta)$ function can be written in the following way:

$$1 - \beta = P\left(\hat{p} > z_{\alpha/2}\sqrt{\frac{p_0(1-p_0)}{n}} + p_0 \mid p \neq p_0\right) + P\left(\hat{p} < -z_{\alpha/2}\sqrt{\frac{p_0(1-p_0)}{n}} + p_0 \mid p \neq p_0\right)$$

$$= P\left(\frac{\hat{p} - p}{\sqrt{\frac{p(1-p)}{n}}} > z_{\alpha/2}\sqrt{\frac{p_0(1-p_0)}{p(1-p)}} + \sqrt{n}\frac{p_0 - p}{\sqrt{p(1-p)}}\right) + P\left(\frac{\hat{p} - p}{\sqrt{\frac{p(1-p)}{n}}} < -z_{\alpha/2}\sqrt{\frac{p_0(1-p_0)}{p(1-p)}} + \sqrt{n}\frac{p_0 - p}{\sqrt{p(1-p)}}\right).$$

Noting that the power is computed under H_a being true and that therefore the expression

$$\frac{\hat{p} - p}{\sqrt{\frac{p(1-p)}{n}}}$$

is distributed as a standard normal, we can compute the power that we have for different values of the sample size n for specified differences $(p - p_0)$. To estimate the sample size needed to detect a specific difference $(p - p_0)$ for given historical rate p_0, statistical power, and α-level, we can use the power equation above, simply solving for n.

Scenario 2: Two-Sample Power Analysis

We want to test, using a two-sided test with an α-level of 5 percent:

$$H_0: p_m - p_c = 0 \text{ versus } H_a: p_m - p_c \neq 0,$$

where p_m is the employment/unemployment rate for the military spouses and p_c is employment/unemployment rate for the civilian spouses.

We collect two simple random samples of military and civilian spouses of size n and m:

$$x_1, x_2, \ldots, x_n \text{ and } y_1, y_2, \ldots, y_m.$$

Given these two samples, we can estimate the employment (or unemployment) rate for both the military and civilian spouses using the sample proportions

$$\hat{p}_m = \frac{\sum_{i=1}^{n} x_i}{n}$$

and

$$\hat{p}_c = \frac{\sum_{i=1}^{m} y_i}{m}.$$

If the sample sizes n and m are large enough, then both sample proportions will be approximately distributed as normal with means p_m and p_c and variances $p_m(1 - p_m)/n$ and $p_c(1 - p_c)/m$. The difference of the sample proportions, again for n and m large, is approximately normal:

$$(\hat{p}_m - \hat{p}_c) \sim N\left((p_m - p_c); \frac{p_m(1 - p_m)}{n} + \frac{p_c(1 - p_c)}{n}\right).$$

We will be rejecting the null hypothesis when the difference between the two sample proportions is large. The α-level will determine how large such difference has to be in order to reject H_0, when H_0 is true:

$$\alpha = P\left(\left|\frac{\hat{p}_m - \hat{p}_c}{\sqrt{\frac{p(1 - p)(n + m)}{nm}}}\right| > z_{\alpha/2}\right),$$

where p is the common value for the military and civilian rate under the null hypothesis.

The power function, in this case, can be written in the following way:

$$1 - \beta = P\left(\hat{p}_m - \hat{p}_c > z_{\alpha/2} \sqrt{\frac{p(1-p)(n+m)}{nm}} \mid p_m \neq p_c \right) + P\left(\hat{p}_m - \hat{p}_c < -z_{\alpha/2} \sqrt{\frac{p(1-p)(n+m)}{nm}} \mid p_m \neq p_c \right)$$

$$1 - \beta = P\left(z > \frac{z_{\alpha/2}\sqrt{\frac{p(1-p)(n+m)}{nm}} - (p_m - p_c)}{\sqrt{\frac{p_m(1-p_m)}{n} + \frac{p_c(1-p_c)}{m}}} \right) + P\left(z < -\frac{z_{\alpha/2}\sqrt{\frac{p(1-p)(n+m)}{nm}} - (p_m - p_c)}{\sqrt{\frac{p_m(1-p_m)}{n} + \frac{p_c(1-p_c)}{m}}} \right).$$

In the formula above,

$$z = \frac{(\hat{p}_m - \hat{p}_c) - (p_m - p_c)}{\sqrt{\frac{p_m(1-p_m)}{n} + \frac{p_c(1-p_c)}{m}}}$$

is approximately a standard normal random variable (for n and m large).

If the two sample sizes are equal ($n = m$), then the power function can be written in the following way:

$$1 - \beta = P\left(z > z_{\alpha/2}\sqrt{\frac{2p(1-p)}{p_m(1-p_m) + p_c(1-p_c)}} - (p_m - p_c)\sqrt{\frac{n}{p_m(1-p_m) + p_c(1-p_c)}} \right) +$$

$$P\left(z < -z_{\alpha/2}\sqrt{\frac{2p(1-p)}{p_m(1-p_m) + p_c(1-p_c)}} - (p_m - p_c)\sqrt{\frac{n}{p_m(1-p_m) + p_c(1-p_c)}} \right).$$

The case of $n = m$ is the case in which the power is the largest. In this study, however, the sample size for the civilian spouses m is going to be much lager than the sample size for the military spouses.

We are going to conduct two power analyses as we conducted for the one-sample case assuming that the sample size for the civilian spouses is $m = 10{,}000$. Note that, because the sample size for the civilian spouses is so large, we could actually consider the p_c known. Making this assumption would reduce this power analysis to the power analysis conducted for the one-sample case. Similarly to the one-sample case we are going to specify differences $(p_m - p_c)$ and determine the sample size n needed to detect such differences with a statistical power of 80 percent.

It is customary to set, in the power formula above, p equal to

$$\frac{np_m + mp_c}{n + m}.$$

Unfortunately, n is not known, because we are using the power formula above to determine the sample size n such that the power equals 80 percent. However since m is likely to be much larger than n, the ratio above is likely to be very close to p_c. Therefore we conducted the analysis setting $p = p_c$. Similar to the one-sample analyses, to estimate an adequate sample size for given level of the civilian spouses' rate (p_c), statistical power, and α-level, we can use the power equation above and solve for n for specific differences ($p_m - p_c$).

Bibliography

Bregger, John E., and Steven E. Haugen, "BLS Introduces New Range of Alternative Unemployment Measures," *Monthly Labor Review,* Vol. 118, No. 10, 1995, pp. 19–26.

Bureau of Labor Statistics, *Current Population Survey: Design and Methodology*, Technical Paper 63RV, March 2002. Online at http://www.bls.census.gov/cps/tp/tp63.htm (as of April 24, 2005).

Bureau of Labor Statistics, *Handbook of Methods*, 2003. Online at http://www.bls.gov/opub/hom/home.htm (as of June 23, 2005).

Bureau of Labor Statistics, *News*, March 2004. Online at http://www.bls.gov/news.release/archives/empsit_04022004.pdf (as of June 23, 2005).

Clogg, Clifford C., Scott R. Eliason, and Kevin T. Leicht, *Analyzing the Labor Force: Concepts, Measures, and Trends*, Kluwer Academic: New York, 2001.

Clogg, Clifford C., Teresa A. Sullivan, and Jan E. Mutchler, "On Measuring Underemployment and Inequality in the Labor Force," *Social Indicators Research*, Vol. 12, 1986, pp. 117–152.

Defense Manpower Data Center, "1999 Survey of Active-Duty Personnel: Administration, Datasets, and Codebook," DMDC Report No. 2000-005, December 2000.

Defense Manpower Data Center, "August 2004 Status of Forces Survey of Active-Duty Members: Administration, Datasets, and Codebook," DMDC Report No. 2004-016, August 2004.

Harrell, Margaret C., Nelson Lim, Laura W. Castaneda, and Daniela Golinelli, *Working Around the Military: Challenges to Military Spouse Employment and Education*, Santa Monica, Calif.: RAND Corporation, MG-196-OSD, 2004. Online at http://www.rand.org/pubs/monographs/MG196/ (as of April 25, 2006).

Hosek, James, Beth Asch, C. Christine Fair, Craig Martin, and Michael Mattock, *Married to the Military: The Employment and Earnings of Military Wives Compared with Those of Civilian Wives*, Santa Monica, Calif.: RAND Corporation, MR-1565-OSD, 2002. Online at http://www.rand.org/pubs/monograph_reports/MR1565 (as of April 25, 2006).

Office of the Deputy Under Secretary of Defense for Military Community and Family Policy, *1st Quadrennial Quality of Life Review Report to Congress,* May 2004. Online at http://www.military-homefront.dod.mil/portal/page?_pageid=73,46096&_dad=itc&_schema=PORTAL&current_id=20.20.60.70.0.0.0.0.0 (as of April 25, 2006).

Scheaffer, Richard L., William Mendenhall III, and R. Lyman Ott, *Elementary Survey Sampling*, 5th ed., Belmont, Calif.: Duxbury Press, 1996.